MEAL WORM FARMING GUIDE

From Setup to Harvest: A Complete Handbook for Sustainable Mealworm Farming

Edwin J. Harrison

Copyright © Edwin J. Harrison, 2024.

All rights reserved. No part of this publication may be reproduced, distributed or transmitted in any form or by any means, including photocopying, recording or other electronic or mechanical methods, without the prior written permission of the publisher, except in the case of brief quotations embodied in critical reviews and certain other noncommercial uses permitted by copyright law.

TABLE OF CONTENT

INTRODUCTION 6
CHAPTER ONE 10
  GETTING STARTED 10
    1.1 Choosing the Right Mealworm Species 10
    1.2 Sourcing Your First Batch of Mealworms 12
    1.3 Essential Equipment and Supplies 14
CHAPTER TWO 18
  SETTING UP YOUR FARM 18
    2.1 Preparing the Habitat 18
    2.2 Environmental Requirements 20
CHAPTER THREE 22
  MEALWORM CARE AND MAINTENANCE 22
    3.1 Feeding Your Mealworms 22
    3.2 Hydration Methods 24
    3.3 Cleaning and Maintenance 25
CHAPTER FOUR 28
  BREEDING MEALWORMS 28
    4.1 Understanding the Breeding Cycle 28
    4.2 Encouraging Mating and Egg Laying 29
    4.3 Managing Eggs and Larvae 30
    4.4 Ensuring Healthy Growth 31
CHAPTER FIVE 33

HARVESTING MEALWORMS 33

    5.1 Timing the Harvest 33

    5.2 Methods of Harvesting 34

    5.3 Post-Harvest Processing and Storage 35

CHAPTER SIX 39

    TROUBLESHOOTING COMMON PROBLEMS 39

    6.1 Identifying and Preventing Pests 39

    6.2 Dealing with Diseases and Health Issues 41

    6.3 Addressing Environmental Issues 42

CHAPTER SEVEN 45

    USES OF MEALWORMS 45

    7.1 As Feed for Pets and Livestock 45

    7.2 In Human Consumption 47

    7.3 Other Applications 48

CHAPTER EIGHT 51

    SCALING UP YOUR OPERATION 51

    8.1 Expanding Your Farm 51

    8.2 Increasing Production Efficiency 52

    8.3 Marketing and Selling Your Mealworms 54

CHAPTER NINE 57

    SUSTAINABLE AND ETHICAL PRACTICES 57

    9.1 Reducing Waste and Recycling 57

    9.2 Ethical Considerations in Mealworm Farming 58

9.3 Future Trends in Mealworm Farming..................60

CHAPTER TEN 62

RESOURCES AND FURTHER READING 62

10.1 Recommended Books and Articles....................62

10.2 Online Forums and Communities......................63

10.3 Suppliers and Equipment Sources 64

CONCLUSION..66

APPENDICES 69

Glossary of terms...69

Frequently Asked Questions (FAQs)........................70

Record-Keeping Templates 71

INTRODUCTION

WELCOME TO MEALWORM FARMING

Welcome to the fascinating world of mealworm farming! Whether you're a novice looking to start a small-scale project or an experienced farmer aiming to expand your operation, this guide will serve as your comprehensive resource for all things mealworm-related. Mealworm farming is an exciting, sustainable, and profitable venture that offers numerous benefits for various applications, from pet food and livestock feed to human consumption and more.

Benefits of Mealworm Farming

Mealworms, the larval stage of the darkling beetle (Tenebrio molitor), are a highly versatile and valuable resource. Here are some key benefits of mealworm farming:

-**Sustainability:** Mealworms are an eco-friendly alternative to traditional protein sources. They require significantly less land, water, and feed compared to conventional livestock, making them an excellent choice for sustainable farming practices.

-**Nutritional Value:** Rich in protein, vitamins, and minerals, mealworms are an excellent nutritional source for pets, livestock, and even humans. Their high protein content and essential nutrients make them an ideal supplement in various diets.

-Economic Potential: With the growing demand for sustainable protein sources, mealworm farming offers lucrative opportunities for entrepreneurs. Whether you're selling live mealworms, dried mealworms, or mealworm-based products, there's a market ready for you to tap into.

-Ease of Farming: Mealworms are relatively easy to farm and maintain. They have a straightforward lifecycle, and with the right setup and care, you can efficiently manage and grow your mealworm population.

Understanding the Lifecycle of Mealworms

To succeed in mealworm farming, it's essential to understand the lifecycle of mealworms. Mealworms undergo four main stages of development: egg, larva (mealworm), pupa, and adult beetle. Each stage has specific requirements and characteristics:

1. Egg: Female beetles lay tiny, white eggs that are difficult to see without close inspection. The eggs hatch in about one to two weeks, depending on environmental conditions.

2. Larva (Mealworm): Upon hatching, mealworms emerge as small, white larvae. They go through multiple molts as they grow, shedding their exoskeleton several times. This stage can last anywhere from a few weeks to several months, depending on temperature and humidity.

3. Pupa: After reaching full size, mealworms enter the pupal stage. They do not eat or move much during this time as they undergo metamorphosis. This stage lasts for about one to three weeks.

4. Adult Beetle: Adult darkling beetles emerge from the pupal stage, ready to mate and lay eggs, starting the cycle anew. Adult beetles can live for several months, continuously contributing to the reproduction of your mealworm colony.

How to Use This Guide

This guide is designed to be your go-to resource for all aspects of mealworm farming. Each chapter covers specific topics, from setting up your farm and caring for mealworms to harvesting and troubleshooting common issues. Whether you're looking for detailed instructions or quick tips, you'll find valuable information to help you succeed in your mealworm farming journey.

We encourage you to read through the guide, take notes, and refer back to sections as needed. With dedication and the right knowledge, you'll be well on your way to running a successful mealworm farm. Happy farming!

CHAPTER ONE

GETTING STARTED

1.1 Choosing the Right Mealworm Species

Mealworms are the larval form of the darkling beetle, and there are several species within this family. However, the most commonly farmed species for both commercial and personal use are the Tenebrio molitor and Zophobas morio.

Tenebrio molitor (Common Mealworm):

- **Size:** These mealworms typically grow to about 2.5 cm (1 inch) in length.
- **Lifecycle:** Their lifecycle is well-suited for continuous farming, with a shorter pupation period and a high rate of reproduction.
- **Usage:** They are widely used for feeding pets such as reptiles, birds, and fish. They are also increasingly popular as a protein source for human consumption due to their high nutritional value.
- **Advantages:** Tenebrio molitor is easy to raise and manage, making them ideal for beginners. They

have a moderate growth rate and do not require very specific conditions to thrive.

Zophobas morio (Superworm):

- **Size:** Superworms are larger, growing up to 5 cm (2 inches) or more.
- **Lifecycle:** They have a longer larval stage and do not pupate as readily as common mealworms unless isolated. This makes them a continuous source of larvae for a longer period.
- **Usage:** Superworms are favored for feeding larger reptiles and amphibians due to their size and higher fat content.
- **Advantages:** They are more active and have a tougher exoskeleton, making them a more stimulating prey item for larger pets. However, they require more space and slightly different care compared to common mealworms.

Choosing the Right Species:

- **Purpose:** Consider the primary purpose of your mealworm farming. If you aim to provide food for smaller pets or human consumption, Tenebrio molitor might be the best choice. For larger pets, Zophobas morio may be more suitable.

- **Experience Level:** Beginners may find Tenebrio molitor easier to manage due to their less demanding care requirements and faster lifecycle.
- **Space and Resources:** Superworms require more space and resources, so ensure you have the capacity to meet their needs if you choose this species.

1.2 Sourcing Your First Batch of Mealworms

Once you've decided on the species, the next step is to source your first batch of mealworms. There are several options for acquiring your initial stock:

1. Local Pet Stores:

- **Convenience:** Pet stores often carry mealworms for reptile and bird owners. This can be a quick and easy way to start.
- **Quality:** Ensure the mealworms are healthy and active. Avoid any that appear lethargic or have visible signs of disease.

2. Online Suppliers:

- **Variety:** Many online suppliers offer different sizes and quantities of mealworms. This can be especially useful if you need a large number or a specific size.
- **Delivery:** Choose a reputable supplier with good reviews to ensure the mealworms are shipped quickly and arrive in good condition.

3. Local Farmers or Breeders:

- **Community:** Connecting with local farmers or breeders can provide valuable insights and support. They may also offer healthier stock, as the mealworms haven't been subjected to the stresses of shipping.
- **Networking:** This option can also help you build a network of local contacts for advice and future sourcing needs.

4. Starting from Wild-Caught Specimens:

- **Caution:** While this method can be cost-effective, it carries the risk of introducing pests and diseases to your farm. It's generally recommended to start with mealworms from a known, reputable source.

1.3 Essential Equipment and Supplies

Setting up a mealworm farm requires some essential equipment and supplies to create a suitable environment for your mealworms to thrive. Here's what you'll need:

1. Containers and Bins:

- **Types:** Plastic bins, glass aquariums, or specialized insect farming containers are all suitable. Ensure the containers are escape-proof and have adequate ventilation.
- **Size:** The size of the container will depend on the scale of your operation. For beginners, a few medium-sized plastic bins (around 20-30 liters) are often sufficient.
- **Quantity:** It's a good idea to have multiple containers to separate different stages of the mealworm lifecycle (larvae, pupae, and beetles).

2. Substrate:

- **Material:** The substrate serves as both bedding and a food source. Common options include wheat bran, oats, or a mixture of both.
- **Depth:** Fill the containers with a layer of substrate about 5-10 cm (2-4 inches) deep. This provides enough space for the mealworms to burrow and thrive.
- **Maintenance:** Regularly check and replace the substrate to keep it clean and free of mold.

3. Food:

- **Primary Diet:** Mealworms feed on the substrate, so ensure it is always available and fresh.
- **Supplementary Foods:** Provide additional nutrition with slices of vegetables (carrots, potatoes) or fruits (apples, cucumbers). These also serve as a moisture source.
- **Frequency:** Replace supplementary foods every few days to prevent spoilage and mold growth.

4. Moisture Source:

- **Hydration:** Mealworms need a moisture source, typically provided by fresh vegetable or fruit slices.

Avoid over-moistening the substrate to prevent mold.
- **Alternatives:** Some farmers use water gel crystals as a moisture source, which can reduce the risk of mold.

5. Heating and Temperature Control:

- **Optimal Temperature:** Mealworms thrive at temperatures between 20-25°C (68-77°F). Use a thermometer to monitor the temperature.
- **Heating Options:** If your environment is too cold, consider using a heat mat or lamp to maintain the optimal temperature range.

6. Ventilation:

- **Airflow:** Proper ventilation is crucial to prevent mold and maintain a healthy environment. Ensure your containers have mesh lids or drilled holes to allow airflow while preventing escapes.
- **Humidity:** Maintain a humidity level of around 50-70%. Too high humidity can lead to mold growth, while too low can cause dehydration.

7. Cleaning Supplies:

- **Tools:** Use small scoops, sifters, and brushes to clean and maintain your mealworm containers.
- **Frequency:** Regularly clean the containers to remove waste, uneaten food, and dead insects to maintain a healthy environment.

8. Record-Keeping:

- **Logs:** Keep detailed records of feeding schedules, cleaning routines, and breeding cycles to optimize your farming practices and ensure consistent production.

By choosing the right species, sourcing healthy mealworms, and equipping yourself with the essential supplies, you'll be well on your way to establishing a successful mealworm farm.

CHAPTER TWO

SETTING UP YOUR FARM

2.1 Preparing the Habitat

Containers and Bins:

Choosing the right containers is crucial for the success of your mealworm farming operation. The containers should be escape-proof, easy to clean, and provide sufficient space for the mealworms to move and grow.

- **Types:** Plastic bins, glass aquariums, or specialized insect farming containers are all suitable options. Plastic bins are often preferred due to their affordability, durability, and ease of handling. Ensure the bins are made of food-grade material to prevent any harmful chemicals from leaching into the substrate.
- **Size:** The size of your containers will depend on the scale of your operation. For small-scale farming, bins with a capacity of 20-30 liters are typically sufficient. As you expand, you may need larger containers or multiple bins.

- **Lids and Ventilation:** Secure lids are essential to prevent mealworms from escaping and to keep out pests. However, proper ventilation is also crucial to maintain airflow and prevent mold. Use lids with mesh inserts or drill small holes in the sides of the bins to allow for adequate ventilation.

Substrate Selection and Preparation:

The substrate serves as both the bedding and primary food source for your mealworms. Choosing the right substrate and preparing it correctly will ensure a healthy environment for your mealworms to thrive.

- **Materials:** Common substrate materials include wheat bran, oats, cornmeal, or a combination of these. Wheat bran is often preferred due to its high nutritional value and ease of handling.
- **Depth:** Fill your containers with a layer of substrate about 5-10 cm (2-4 inches) deep. This provides enough space for the mealworms to burrow and thrive.
- **Preparation:** Before adding the substrate to the bins, sift it to remove any large particles or

contaminants. Ensure the substrate is dry to prevent mold growth. Regularly check and replace the substrate to keep it clean and free of mold.

2.2 Environmental Requirements

Creating the optimal environment for your mealworms involves maintaining the right temperature, humidity, and lighting conditions. These factors significantly impact their growth, reproduction, and overall health.

Temperature and Humidity:

- **Optimal Temperature:** Mealworms thrive at temperatures between 20-25°C (68-77°F). Maintaining a stable temperature within this range is crucial for their growth and reproduction. Use a thermometer to monitor the temperature regularly.
- **Heating Options:** If your environment is too cold, consider using a heat mat or lamp to maintain the optimal temperature range. Avoid placing the bins directly on cold floors or in drafty areas.
- **Humidity:** Maintain a humidity level of around 50-70%. Too high humidity can lead to mold growth,

while too low can cause dehydration. Use a hygrometer to monitor humidity levels. To control humidity, you can add or remove moisture sources (such as vegetable slices) or adjust ventilation.

Light and Darkness:

- **Lighting:** Mealworms do not require special lighting conditions. They can thrive in natural light or low artificial light environments. Avoid placing the bins in direct sunlight, as it can cause overheating and stress to the mealworms.
- **Darkness:** While mealworms can tolerate some light, they generally prefer darker conditions. Providing some cover, such as a cloth or cardboard over the bins, can help create a more favorable environment.

CHAPTER THREE

MEALWORM CARE AND MAINTENANCE

3.1 Feeding Your Mealworms

Nutritional Requirements:

Mealworms have specific nutritional needs that must be met to ensure their healthy growth and development. Their diet consists primarily of the substrate they live in, supplemented with various food items to provide a balanced nutrition.

- **Primary Diet:** The main component of the mealworms' diet is the substrate, which typically includes wheat bran, oats, or cornmeal. These grains provide essential carbohydrates and some proteins needed for their growth.
- **Protein:** Mealworms require protein for growth and development. In addition to their primary diet, you can add protein-rich foods such as powdered milk, ground dog food, or fish food flakes.
- **Vitamins and Minerals:** Adding fresh vegetables and fruits like carrots, potatoes, apples, and leafy

greens can supply necessary vitamins and minerals. These also help keep the substrate moist, providing both nutrition and hydration.
- **Calcium:** Adding a small amount of powdered calcium supplement to the substrate can help with the development of a strong exoskeleton.

Feeding Schedule and Options:

Maintaining a consistent feeding schedule ensures that your mealworms receive a steady supply of nutrients without overfeeding, which can lead to mold growth and waste accumulation.

- **Daily Feeding:** Check the food supply daily. Add fresh slices of vegetables or fruits every 1-2 days to provide moisture and additional nutrients.
- **Weekly Feeding:** Once a week, replenish the substrate if it appears to be depleted. Mix in additional wheat bran, oats, or other grain-based substrates.
- **Avoid Overfeeding:** Overfeeding can lead to food spoilage and mold growth, which can harm your

mealworms. Add only what they can consume within a day or two.

- **Rotation:** Rotate the types of fresh foods you provide to ensure a balanced intake of nutrients. For example, alternate between carrots, apples, and leafy greens.

3.2 Hydration Methods

Proper hydration is crucial for the health and growth of your mealworms. While they get most of their moisture from their food, providing additional hydration sources is essential.

- **Fresh Vegetables and Fruits:** The easiest and most effective way to hydrate your mealworms is by adding slices of fresh vegetables and fruits. Carrots, potatoes, apples, and cucumbers are excellent choices as they provide moisture and nutrition.
- **Water Gel Crystals:** Water gel crystals can be used as a supplemental hydration source. These crystals absorb water and slowly release it, reducing the risk of mold compared to using water dishes.

- **Avoid Water Dishes:** Direct water sources such as bowls or dishes should be avoided as mealworms can drown, and the water can quickly become a breeding ground for bacteria and mold.

Tips for Hydration:

- **Replace Regularly:** Check the hydration sources daily and replace them every 1-2 days to prevent spoilage and mold.
- **Quantity:** Provide enough hydration sources to keep the substrate slightly moist but not wet. The substrate should be damp to the touch but not soggy.

3.3 Cleaning and Maintenance

Regular cleaning and maintenance of your mealworm farm are essential to prevent disease, mold growth, and to ensure a healthy environment for your mealworms.

Cleaning Schedule:

- **Daily Checks:** Perform daily checks to remove any uneaten food, moldy items, or dead mealworms.

This helps maintain a clean and healthy environment.

- **Weekly Cleaning:** Once a week, sift through the substrate to remove waste (frass) and dead insects. A fine mesh sieve can be helpful for this task. Replace or replenish the substrate as needed.
- **Monthly Deep Cleaning:** Every month, perform a more thorough cleaning. Transfer the mealworms to a temporary container, dispose of the old substrate, and clean the bins with mild soap and water. Ensure the bins are completely dry before adding fresh substrate and returning the mealworms.

Maintenance Tips:

- **Preventing Mold:** Keep the substrate dry by providing appropriate ventilation and avoiding overfeeding. Remove any moldy food immediately to prevent it from spreading.
- **Pest Control:** Regularly check for pests such as mites or beetles. If you notice any, take immediate action to remove them and prevent infestation. Cleaning and maintaining a dry environment can help deter pests.

- **Record Keeping:** Keep detailed records of feeding schedules, cleaning routines, and any issues that arise. This helps track the health of your mealworm population and identify patterns that may need addressing.

CHAPTER FOUR

BREEDING MEALWORMS

4.1 Understanding the Breeding Cycle

Understanding the breeding cycle of mealworms is crucial for successful farming. Mealworms undergo complete metamorphosis, progressing through four stages: egg, larva, pupa, and adult beetle.

- **Egg Stage:** Female beetles lay tiny, white eggs in the substrate. These eggs are almost microscopic, making them difficult to see without close inspection. The incubation period for eggs is typically 4-19 days, depending on environmental conditions.
- **Larva Stage:** Upon hatching, mealworms emerge as small larvae. This is the primary growth phase, where they shed their exoskeleton multiple times, a process known as molting. The larval stage lasts for about 10 weeks but can vary based on temperature, humidity, and food availability.

- **Pupa Stage:** After the larval stage, mealworms transform into pupae. This stage is a resting phase where the larvae metamorphose into adult beetles. The pupal stage lasts approximately 6-18 days.
- **Adult Beetle Stage:** Adult darkling beetles emerge from the pupae. They are initially soft and white but harden and darken within a few hours. Adult beetles are responsible for mating and laying eggs. They live for several months, continually contributing to the reproduction cycle.

4.2 Encouraging Mating and Egg Laying

Creating optimal conditions for mating and egg laying is essential for maintaining a productive mealworm farm.

- **Separation of Adults:** To maximize breeding efficiency, it's beneficial to separate adult beetles from larvae and pupae. This prevents the adults from accidentally consuming the eggs and provides a dedicated space for mating and egg laying.
- **Environmental Conditions:** Maintain an optimal temperature range of 25-27°C (77-81°F) and a

humidity level of around 50-70%. These conditions encourage mating behavior and egg production.

- **Substrate Depth:** Provide a deep layer of substrate (5-10 cm) in the breeding container. This gives beetles ample space to lay eggs and helps protect the eggs from disturbance.
- **Feeding:** Ensure a consistent supply of nutritious food and moisture. A well-fed beetle population is more likely to reproduce successfully.
- **Light and Darkness:** While beetles can tolerate light, they generally prefer dim conditions for mating and egg laying. Avoid direct sunlight and consider covering the containers to create a darker environment.

4.3 Managing Eggs and Larvae

Proper management of eggs and larvae is crucial to maintain a healthy and productive mealworm population.

- **Egg Care:** After the eggs are laid, it's essential to handle the substrate gently to avoid damaging them. Avoid sifting or disturbing the substrate excessively.

- **Larval Separation:** Once the eggs hatch and larvae emerge, it's helpful to separate them from the adult beetles to prevent accidental predation and to create optimal growth conditions for the larvae.
- **Substrate Maintenance:** Regularly check and replenish the substrate to provide a clean and nutritious environment for the larvae. Remove any moldy or spoiled food to prevent contamination.
- **Feeding Larvae:** Provide a consistent supply of food and moisture. Fresh vegetables and fruits are essential for hydration, while the substrate should be nutritious to support larval growth.

4.4 Ensuring Healthy Growth

Ensuring healthy growth of mealworms requires consistent attention to their environment, diet, and overall care.

- **Optimal Temperature and Humidity:** Maintain stable environmental conditions to support healthy growth. Use a thermometer and hygrometer to monitor and adjust temperature and humidity levels as needed.

- **Adequate Space:** Avoid overcrowding by providing sufficient space in the containers. Overcrowding can lead to competition for food and resources, increasing stress and reducing growth rates.
- **Regular Cleaning:** Keep the habitat clean by regularly removing waste (frass), uneaten food, and dead insects. A clean environment reduces the risk of disease and promotes healthier mealworms.
- **Monitoring Health:** Regularly inspect your mealworms for signs of disease or stress. Healthy mealworms are active and have a uniform color. Any abnormalities, such as discoloration or lethargy, should be addressed promptly.
- **Feeding and Hydration:** Ensure a balanced diet with adequate protein, vitamins, and minerals. Regularly provide fresh food for hydration and nutritional supplementation.

CHAPTER FIVE

HARVESTING MEALWORMS

5.1 Timing the Harvest

Timing your mealworm harvest correctly is essential for ensuring the highest quality and nutritional value of the mealworms.

- **Optimal Size:** Mealworms are typically harvested when they reach their maximum size, which is usually about 2.5 cm (1 inch) for common mealworms (Tenebrio molitor) and up to 5 cm (2 inches) for superworms (Zophobas morio). This stage is just before they enter the pupal stage.
- **Lifecycle Consideration:** Monitor the lifecycle closely. Harvesting should occur when the mealworms are still active larvae and have not yet transitioned to pupae. Once they start to pupate, their nutritional value and usability as feed decrease.
- **Batch Management:** To ensure a continuous supply, manage your mealworm farm in batches.

This way, you can stagger the harvest times and have a steady production cycle without depleting your entire population at once.

5.2 Methods of Harvesting

There are several methods for harvesting mealworms, each suited to different scales of farming and specific needs.

1. Manual Harvesting:

- **Process:** Use a small scoop or sieve to separate mealworms from the substrate. Gently sift the substrate to collect the mealworms while allowing the frass and smaller particles to fall through.
- **Advantages:** This method is simple and requires minimal equipment. It is ideal for small-scale operations or hobbyists.
- **Disadvantages:** Manual harvesting can be time-consuming and labor-intensive, especially for larger farms.

2. Automated Harvesting:

- **Equipment:** Use specially designed mealworm sifters or automated harvesting machines. These devices use vibration or rotation to separate mealworms from the substrate efficiently.
- **Advantages:** Automated systems significantly reduce labor and time required for harvesting. They are ideal for larger-scale operations.
- **Disadvantages:** The initial cost of automated equipment can be high. It may also require regular maintenance to ensure proper functioning.

3. Light and Heat Traps:

- **Process:** Mealworms are attracted to heat and light. Set up light or heat sources to draw the mealworms to specific areas, making them easier to collect.
- **Advantages:** This method can be useful for collecting large quantities of mealworms quickly.
- **Disadvantages:** It requires careful setup and monitoring to ensure that the mealworms are not exposed to excessive heat or light, which can harm them.

5.3 Post-Harvest Processing and Storage

Proper post-harvest processing and storage are crucial to maintaining the quality and longevity of your harvested mealworms.

1. Cleaning:

- **Initial Cleaning:** After harvesting, place the mealworms in a clean container. Remove any remaining substrate, frass, or debris by gently sifting them.
- **Final Rinse:** For human consumption or certain pet uses, rinse the mealworms in clean water. This helps remove any residual substrate or contaminants. Ensure the mealworms are completely dry before proceeding to the next step to prevent mold growth during storage.

2. Drying:

- **Air Drying:** Spread the cleaned mealworms in a single layer on a mesh or perforated tray. Allow them to air dry in a well-ventilated area. This method is suitable for small batches and warm, dry climates.

- **Oven Drying:** For faster results, use an oven set at a low temperature (around 60-70°C or 140-160°F). Spread the mealworms on a baking sheet and dry them for several hours until they are completely dehydrated.
- **Dehydrator:** A food dehydrator can also be used for consistent and efficient drying. Follow the manufacturer's instructions for optimal results.

3. Freezing:

- **Process:** Place the cleaned and dried mealworms in airtight containers or freezer bags. Label the containers with the date of harvest and store them in the freezer.
- **Advantages:** Freezing preserves the mealworms for extended periods, preventing spoilage and retaining nutritional value. Frozen mealworms can be used as needed and defrosted in small batches.

4. Storing:

- **Short-Term Storage:** For mealworms that will be used within a few weeks, store them in a cool, dry place in airtight containers. Ensure they are kept away from direct sunlight and moisture to prevent mold growth.

- **Long-Term Storage:** For long-term storage, freezing or drying is recommended. Properly dried mealworms can be stored in sealed containers at room temperature for several months. Ensure the containers are kept in a dry and cool environment to maintain quality.

CHAPTER SIX

TROUBLESHOOTING COMMON PROBLEMS

6.1 Identifying and Preventing Pests

Pests can pose a significant threat to your mealworm farming operation, potentially causing damage to your stock and disrupting production. Identifying and preventing pests is essential for maintaining a healthy mealworm population.

Common Pests:

- **Mites:** Tiny, often red or black pests that can infest mealworm colonies, especially in humid conditions. They feed on mealworms and their eggs.
- **Flour Beetles:** Small beetles that may infest stored grains and mealworm substrate, competing for food and space with mealworms.
- **Ants:** Ants are attracted to mealworms and their food sources, potentially disrupting colonies and causing stress to mealworms.

Preventive Measures:

- **Cleanliness:** Maintain a clean and hygienic environment. Regularly remove debris, uneaten food, and frass (waste) to reduce potential food sources for pests.
- **Ventilation:** Ensure proper ventilation in your mealworm containers to discourage mold growth, which can attract pests like mites.
- **Screening:** Use fine mesh screens or covers on containers to prevent pests like ants from accessing mealworm colonies.
- **Monitoring:** Regularly inspect your mealworm colonies for signs of pests. Early detection allows for prompt action to prevent infestations from spreading.

Control Measures:

- **Natural Predators:** Introduce natural predators of pests, such as predatory mites, to help control pest populations without harming mealworms.
- **Traps:** Use sticky traps or bait stations to catch ants and beetles. Place these traps near mealworm containers but away from the mealworms themselves to avoid accidental trapping.

6.2 Dealing with Diseases and Health Issues

Maintaining the health of your mealworms is crucial for their productivity and longevity. Recognizing common diseases and health issues and taking appropriate action can prevent outbreaks and minimize their impact.

Common Diseases:

- **Mold:** Mold can develop on damp substrate, potentially harming mealworms and contaminating food sources. Ensure substrate is kept dry and well-ventilated.
- **Bacterial Infections:** Symptoms may include lethargy, discoloration, or unusual behavior. Maintain cleanliness and hygiene to minimize the risk of bacterial infections.
- **Parasitic Infections:** Internal parasites can affect mealworms, causing weight loss and decreased activity. Proper hygiene and monitoring can help prevent parasitic infections.

Preventive Measures:

- **Clean Environment:** Regularly clean and sanitize mealworm containers, removing any moldy substrate or debris.
- **Quarantine:** Quarantine new mealworms or contaminated individuals to prevent the spread of diseases.
- **Hygiene Practices:** Wash hands and use clean equipment when handling mealworms to avoid introducing contaminants.

Treatment Options:

- **Substrate Replacement:** Remove and replace contaminated substrate immediately to prevent the spread of mold or bacteria.
- **Isolation:** Separate infected mealworms from the main colony to prevent further contamination. Monitor isolated individuals closely and provide supportive care if necessary.

6.3 Addressing Environmental Issues

Environmental factors such as temperature, humidity, and light can impact the health and productivity of your mealworms. Addressing these issues ensures optimal conditions for growth and reproduction.

Temperature and Humidity:

- **Maintain Stability:** Fluctuations in temperature and humidity can stress mealworms and affect their growth. Use thermometers and hygrometers to monitor and maintain stable environmental conditions.
- **Optimal Range:** Mealworms thrive at temperatures between 20-25°C (68-77°F) and humidity levels around 50-70%. Adjust heating or cooling methods as needed to keep within these ranges.

Lighting:

- **Light Exposure:** Mealworms prefer dim or low-light conditions. Avoid exposing them to direct sunlight, as it can cause overheating and stress.
- **Light Cycles:** Maintain consistent light cycles, mimicking natural day and night patterns. This helps regulate mealworm activity and reproduction.

Ventilation:

- **Airflow:** Proper ventilation is essential to prevent mold growth and ensure a healthy environment. Use containers with ventilation holes or mesh lids to promote airflow without allowing pests to enter.

CHAPTER SEVEN

USES OF MEALWORMS

7.1 As Feed for Pets and Livestock

Mealworms are a nutritious and versatile source of protein and essential nutrients, making them suitable for feeding various pets and livestock.

Benefits:

- **High Protein Content:** Mealworms are rich in protein, essential amino acids, and fats, making them an excellent dietary supplement for animals requiring high-protein diets.
- **Nutritional Balance:** They provide essential vitamins (such as B vitamins) and minerals (like iron and calcium), contributing to overall health and vitality.
- **Palatability:** Many animals find mealworms palatable, making them an attractive option for picky eaters or animals needing encouragement to eat.

Common Uses:

- **Reptiles and Amphibians:** Many reptiles, such as bearded dragons, leopard geckos, and amphibians like frogs and toads, thrive on a diet that includes mealworms. They provide a natural prey item that supports their dietary needs.
- **Birds:** Certain bird species, including chickens, ducks, and parrots, enjoy mealworms as a treat or as part of a balanced diet. They provide essential nutrients that support feather health and egg production.
- **Small Mammals:** Some small mammals, such as hedgehogs and sugar gliders, benefit from mealworms as a protein-rich supplement to their diet.

Feeding Tips:

- **Variety:** Offer mealworms as part of a varied diet to ensure nutritional balance and prevent dietary deficiencies.

- **Supplementation:** Use mealworms as a supplement rather than a primary food source, particularly for animals requiring a more diverse diet.

7.2 In Human Consumption

Mealworms are gaining popularity as a sustainable and nutritious source of protein for human consumption, known as entomophagy.

Nutritional Value:

- **Protein-Rich:** Mealworms contain high-quality protein comparable to traditional meat sources.
- **Essential Nutrients:** They are rich in vitamins (such as B vitamins) and minerals (like iron and zinc), essential for human health.
- **Sustainable:** Compared to conventional livestock, mealworms require significantly less water, feed, and space, making them a more sustainable protein source.

Culinary Uses:

- **Cooking:** Mealworms can be cooked in various ways, such as roasted, fried, or baked, to enhance their flavor and texture.
- **Ingredients:** They can be ground into flour for baking or used whole in dishes like salads, stir-fries, or as a protein topping.

Considerations:

- **Safety:** Ensure mealworms are sourced from reputable suppliers and properly prepared to minimize the risk of contamination.
- **Allergies:** Individuals with shellfish or insect allergies should exercise caution when consuming mealworms.

7.3 Other Applications

Mealworms have diverse applications beyond feed and human consumption, leveraging their nutritional content and environmental benefits.

Environmental Benefits:

- **Biodegradation:** Mealworms can break down organic waste, such as food scraps and agricultural by-products, into nutrient-rich compost.
- **Bioconversion:** They can convert organic waste into protein-rich biomass, potentially used for animal feed or biofuel production.
- **Research:** Mealworms are used in scientific research, studying topics like nutrition, ecology, and waste management.

Industrial Uses:

- **Bioplastics:** Mealworms have been explored for their potential in producing biodegradable plastics from chitin extracted from their exoskeleton.
- **Medical Applications:** Compounds found in mealworms are being investigated for potential pharmaceutical applications, such as antimicrobial agents.

Educational and Hobbyist Uses:

- **Educational Tools:** Mealworms are often used in classrooms and educational settings to teach biology, ecology, and life cycles.

- **Hobby Farming:** Hobbyists raise mealworms as pets or for personal use, such as feeding their own pets or experimenting with sustainable practices.

CHAPTER EIGHT

SCALING UP YOUR OPERATION

8.1 Expanding Your Farm

Expanding your mealworm farm involves increasing the scale of production to meet growing demand or personal goals. Whether you're scaling up for commercial purposes or expanding as a hobbyist, careful planning and management are essential.

Planning for Expansion:

- **Assess Demand:** Determine the market demand for mealworms in your target area or niche. Research potential customers, such as pet stores, reptile owners, or local farmers.
- **Infrastructure:** Evaluate your current setup and consider additional space, containers, and equipment needed to accommodate increased production.
- **Scaling Up:** Gradually increase the number of mealworm colonies or batches to avoid

overwhelming yourself and maintain control over quality and productivity.

Financial Considerations:

- **Budgeting:** Estimate the costs associated with expansion, including equipment, supplies, and operational expenses.
- **Investment:** Consider sources of funding or investment opportunities if significant capital is required for expansion.

Regulatory Compliance:

- **Permits and Regulations:** Depending on your location and scale of operation, you may need permits or licenses to legally sell mealworms or operate a farming business. Research local regulations and ensure compliance.

8.2 Increasing Production Efficiency

Improving production efficiency ensures that your mealworm farm operates smoothly and maximizes output without compromising quality.

Efficiency Strategies:

- **Optimized Feeding:** Develop a feeding schedule that meets the nutritional needs of mealworms without overfeeding, which can lead to waste and mold.
- **Automated Systems:** Invest in automation, such as automated feeders or harvesting equipment, to streamline tasks and reduce labor costs.
- **Temperature Control:** Maintain stable environmental conditions to promote consistent growth and development.
- **Monitoring:** Use monitoring tools, such as temperature and humidity sensors, to track conditions and make timely adjustments.
- **Batch Management:** Implement batch management strategies to stagger production cycles and maintain a steady supply of mealworms.

Quality Control:

- **Regular Inspections:** Conduct regular inspections of mealworm colonies for signs of pests, diseases, or environmental stress. Address issues promptly to prevent disruptions in production.

- **Training and Education:** Continuously train yourself and any staff on best practices for mealworm farming to improve efficiency and maintain high standards.

8.3 Marketing and Selling Your Mealworms

Effectively marketing and selling your mealworms is crucial for reaching your target audience and generating revenue from your farm.

Identifying Your Market:

- **Target Audience:** Identify potential customers, such as pet owners, reptile enthusiasts, bird breeders, or sustainable food advocates.
- **Value Proposition:** Highlight the nutritional benefits, sustainability, and quality of your mealworms to differentiate them from competitors.

Marketing Strategies:

- **Online Presence:** Create a professional website or social media profiles to showcase your mealworm products, farming practices, and customer testimonials.
- **Networking:** Attend local pet expos, farmers markets, or trade shows to promote your mealworms and connect with potential buyers.
- **Word-of-Mouth:** Encourage satisfied customers to spread the word about your mealworms through reviews and referrals.

Sales Channels:

- **Direct Sales:** Sell mealworms directly to consumers through your website, local markets, or farm stand.
- **Wholesale:** Partner with pet stores, feed suppliers, or distributors to sell mealworms in bulk.
- **Online Platforms:** List your mealworms on online marketplaces or platforms specializing in pet supplies or live feed.

Customer Service:

- **Quality Assurance:** Ensure consistent product quality and customer satisfaction through reliable delivery and responsive customer service.
- **Feedback:** Collect feedback from customers to continually improve your products and services.

CHAPTER NINE

SUSTAINABLE AND ETHICAL PRACTICES

9.1 Reducing Waste and Recycling

Reducing waste and promoting recycling are essential practices in sustainable mealworm farming, benefiting both the environment and operational efficiency.

Waste Reduction Strategies:

- **Efficient Feeding:** Optimize feeding practices to minimize excess food that could lead to waste and mold. Provide only what mealworms can consume within a reasonable time.
- **Substrate Management:** Use renewable and biodegradable substrates, such as wheat bran or oats, which can be composted after use.
- **Composting:** Utilize mealworm frass (waste) as nutrient-rich compost for gardening or agricultural purposes.
- **Recycling Containers:** Use durable and reusable containers for housing mealworm colonies, reducing the need for disposable materials.

Recycling Practices:

- **Organic Waste:** Process organic waste, such as vegetable scraps or spent grains, as feed for mealworms, promoting a closed-loop system.
- **Packaging:** Use recyclable or biodegradable packaging materials for selling mealworms to minimize environmental impact.
- **Water Conservation:** Implement water-saving techniques, such as drip irrigation or rainwater harvesting, to reduce water consumption in farming operations.

9.2 Ethical Considerations in Mealworm Farming

Ethical considerations in mealworm farming involve ensuring humane treatment of mealworms and addressing broader ethical implications of insect farming.

Humane Treatment:

- **Handling Practices:** Handle mealworms gently and minimize stress during handling and harvesting.

- **Euthanasia:** Use humane methods for euthanasia when necessary, such as freezing, to minimize suffering.
- **Living Conditions:** Provide adequate space, proper nutrition, and suitable environmental conditions for mealworms' health and well-being.

Environmental Impact:

- **Sustainability:** Adopt sustainable farming practices that minimize environmental impact, such as reducing resource consumption and greenhouse gas emissions.
- **Biodiversity:** Consider the impact of mealworm farming on local ecosystems and biodiversity, striving to maintain ecological balance.

Consumer Perception:

- **Education:** Educate consumers about the nutritional benefits and sustainability of mealworms to promote acceptance and understanding of insect-based protein sources.

- **Transparency:** Provide transparent information about farming practices and ethical considerations to build trust with consumers.

9.3 Future Trends in Mealworm Farming

The future of mealworm farming is shaped by technological advancements, sustainability goals, and evolving consumer preferences for alternative protein sources.

Technological Innovations:

- **Automation:** Increased use of automated systems for feeding, harvesting, and monitoring to improve efficiency and reduce labor costs.
- **Biotechnology:** Advances in biotechnology for enhancing mealworm nutrition, disease resistance, and growth rates.
- **Smart Farming:** Integration of IoT (Internet of Things) devices and data analytics for real-time monitoring and decision-making in farming operations.

Sustainability Initiatives:

- **Circular Economy:** Emphasis on closed-loop systems where waste from one process becomes input for another, promoting resource efficiency.
- **Carbon Footprint:** Reduction strategies to minimize greenhouse gas emissions associated with mealworm farming through improved energy efficiency and sustainable practices.

Market Expansion:

- **Global Demand:** Increasing adoption of insect-based protein in diverse markets, including animal feed, human nutrition, and biotechnology.
- **Regulatory Landscape:** Development of regulatory frameworks to support the growth of insect farming as a sustainable food source.

Consumer Trends:

- **Health and Nutrition:** Growing awareness of the nutritional benefits of mealworms, such as high protein content and essential nutrients.
- **Environmental Awareness:** Consumer preference for sustainable and eco-friendly food choices, driving demand for insect-based products.

CHAPTER TEN

RESOURCES AND FURTHER READING

10.1 Recommended Books and Articles

Expand your knowledge of mealworm farming with these recommended books and articles that cover various aspects of raising mealworms:

Books:

- *Mealworms: Raise Them, Watch Them, See Them Change* by Adrienne Mason
- *The Complete Guide to Raising Mealworms: Healthy Feeder Insects for Your Reptiles, Birds, and More* by Michelle Ledet Henley
- *Insects as Sustainable Food Ingredients: Production, Processing, and Food Applications* edited by Aaron T. Dossey, Juan A. Morales-Ramos, and M. Guadalupe Rojas

Articles:

- "Mealworm Farming for Beginners: A Step-by-Step Guide" - Available on Backyard Boss
- "Nutritional Value of Mealworms for Animal Feed" - Published in Journal of Insects as Food and Feed
- "Sustainability of Insect Farming for Animal Feed" - Published in Frontiers in Sustainable Food Systems

10.2 Online Forums and Communities

Join online forums and communities to connect with other mealworm farmers, share experiences, and gain insights into best practices:

Forums:

- **BackYard Chickens Forum:** Discuss mealworms as chicken feed and interact with poultry owners.
- **Bearded Dragon.org Forum:** Engage with reptile owners who use mealworms as part of their pet's diet.
- **Mealworm Farming Facebook Group:** Join a community of mealworm enthusiasts sharing tips and experiences.

Communities:

- **Reddit - r/Mealworms:** A subreddit dedicated to mealworm farming, where users discuss everything from breeding tips to troubleshooting.
- **InsectNet Forum:** A community for insect enthusiasts and farmers discussing various insect farming techniques, including mealworms.

10.3 Suppliers and Equipment Sources

Find reliable suppliers and equipment sources for your mealworm farming needs:

Suppliers:

- **Rainbow Mealworms:** Offers live mealworms, superworms, and related products for feeding pets and livestock.
- **GrubTubs:** Provides sustainable insect farming solutions, including mealworms and equipment for hobbyists and commercial farms.
- **Josh's Frogs:** Supplies live insects, including mealworms, for feeding reptiles, amphibians, and other pets.

Equipment Sources:

- **FarmTek:** Offers a range of agricultural and farming equipment, including containers and heating solutions suitable for mealworm farming.
- **Entomo Farms:** Provides equipment and consulting services for insect farming, supporting sustainable production practices.
- **Amazon:** Explore various suppliers on Amazon for mealworm farming equipment, such as containers, feeders, and temperature controllers.

CONCLUSION

Mealworm farming offers a sustainable and versatile solution for meeting various needs, from pet nutrition to human consumption and beyond. Throughout this guide, we've explored the fundamentals of starting and maintaining a successful mealworm farm, covering everything from setup and care to harvesting and marketing.

By embarking on this journey, you've learned how to:

- **Establish Your Farm:** Set up an efficient mealworm farming operation, from choosing the right species and sourcing initial stock to selecting suitable equipment and creating optimal habitats.
- **Manage Carefully:** Understand the nutritional needs, environmental requirements, and lifecycle stages of mealworms to ensure their health and productivity.
- **Harvest and Utilize:** Time your harvests effectively, employ proper harvesting techniques, and process mealworms for various uses, including

pet feed, human consumption, and other applications.
- **Sustainably Farm:** Implement practices to reduce waste, recycle resources, and consider ethical considerations in mealworm farming, ensuring responsible stewardship of resources and respect for animal welfare.

Looking ahead, the future of mealworm farming holds promise with ongoing advancements in technology, growing awareness of sustainability, and increasing acceptance of insect-based protein sources. By staying informed through recommended books, articles, online communities, and connecting with reliable suppliers, you can continue to refine your skills and contribute to the sustainable food production landscape.

Whether you're a hobbyist exploring a new passion or an entrepreneur seeking to meet market demands sustainably, mealworm farming offers opportunities for innovation, growth, and environmental stewardship. Embrace the knowledge gained from this guide and continue to explore, experiment, and contribute to the thriving field of mealworm farming. Together, we can shape a future where sustainable practices and alternative protein sources play a vital role in feeding our world.

APPENDICES

Glossary of terms

1. Substrate: The material used as bedding for mealworms, providing a medium for them to burrow and lay eggs. Common substrates include wheat bran, oats, or a mixture of grains.

2. Frass: The excrement or waste produced by mealworms, which can be used as nutrient-rich compost for plants.

3. Pupa: The stage in the mealworm life cycle where larvae transform into pupae before emerging as adult beetles.

4. Larvae: The immature stage of mealworms, which hatch from eggs and grow until they pupate.

5. Colony: A group of mealworms kept together for breeding, feeding, or harvesting purposes.

6. Ventilation: The process of providing airflow and fresh air exchange within mealworm containers or farms to maintain optimal conditions.

7. Humidity: The level of moisture or water vapor present in the air or substrate, crucial for mealworm growth and development.

8. Incubation: The period during which mealworm eggs are kept in controlled conditions to facilitate hatching.

9. Molting: The process in which mealworms shed their exoskeletons as they grow, allowing for continued growth and development.

10. Cannibalism: Behavior observed in mealworms where they may consume weaker or injured individuals, often a result of overcrowding or stress.

Frequently Asked Questions (FAQs)

Q1: What do mealworms eat? Mealworms primarily feed on grains, such as wheat bran or oats, supplemented with fruits and vegetables for hydration and additional nutrients.

Q2: How do I prevent mold in my mealworm containers? To prevent mold, ensure the substrate is not overly damp and provide adequate ventilation. Regularly remove uneaten food and clean containers to maintain a dry environment.

Q3: How often should I clean mealworm containers? Clean mealworm containers regularly to remove frass, uneaten food, and moldy substrate. Frequency depends on colony size and cleanliness needs but typically ranges from weekly to monthly.

Q4: Can I breed mealworms year-round? Yes, mealworms can be bred year-round with consistent environmental conditions, including temperature and light cycles mimicking their natural habitat.

Q5: How do I harvest mealworms efficiently? Harvest mealworms by separating them from substrate using a sieve or shaking method. Transfer harvested mealworms to a clean container for processing or storage.

Record-Keeping Templates

1. Feeding and Maintenance Log:

- **Date:** _____

- **Feeding Details:** Type of food given, quantity, any observations.
- **Maintenance:** Cleaning details, substrate changes, environmental conditions.

2. Growth and Development Chart:

- **Date:** _____
- **Stage:** Larvae, pupae, adult beetle.
- **Quantity:** Number of mealworms at each stage.
- **Observations:** Any abnormalities or notable changes.

3. Harvest and Yield Record:

- **Date:** _____
- **Harvest Details:** Quantity of mealworms harvested, method used.
- **Yield:** Weight or volume of harvested mealworms.
- **Notes:** Any observations on harvest quality or issues encountered.

4. Environmental Monitoring Template:

- **Date:** _____

- **Temperature:** _____ (°C/°F)
- **Humidity:** _____ (%)
- **Light Exposure:** _____ (hours of light/day)
- **Ventilation:** _____ (frequency of air exchange)

These templates can be customized and used to track key metrics, observations, and activities related to mealworm farming, aiding in management and decision-making for optimal farm performance.

THANKS FOR READING

www.ingramcontent.com/pod-product-compliance
Lightning Source LLC
Chambersburg PA
CBHW071954210526
45479CB00003B/932